Arduino

A Complete Step by Step Guide

Disclaimer

No part of this eBook may be transmitted or shared in any form including print, electronic, photocopy, scanning, mechanical or recording without prior written permission from the author.

The information contained in this eBook is solely for educational purposes.

The author has made all efforts to ensure the information contained in here is authentic and useful. However, the reader is advised to follow the guidelines as their own responsibility. The author is not liable for any personal or commercial damage occurring due to the misinterpretation of the information contained herein. The reader is advised to seek professional help as and when required. The author cannot be pressed for any charges in any circumstances.

Summary

This is the era of automation. It is possible to alter the physical environment automatically through the use of computerized chips and protocols. The most basic use of this technology can be seen in the form of automated lights – the luminosity automatically turns on when a specific measurable level of darkness is crossed. At any level of luminosity above this, the lamps remain off.

With Arduino, this has been made infinitely easier. Arduino produces single board microcontrollers which serve the purpose of a moderator/controller. Contained in the following pages is a detailed guide pertaining to the wonders of a microcontroller. It also covers the ways in which you can personalize it according to your specific needs. Keep reading to find out everything about Arduino!

Contents

Disclaimer 2

Summary 3

Introduction to Arduino 8

    What is Arduino? 10

    Uses of Arduino 13

Arduino Products 16

    Boards 17

        Arduino Uno 17

        Arduino Leonardo 20

        Arduino Due 22

        Arduino Yun 23

        Arduino Micro 25

        Arduino Robot 27

        Arduino Esplora 28

        Arduino ADK 29

        Arduino Ethernet 32

        Arduino Mega 2560 33

        Arduino Mini 35

        LilyPad Arduino USB 36

LilyPad Arduino Simple 38

LilyPad Arduino SimpleSnap 39

LilyPad Arduino 41

Arduino Nano 43

Arduino Pro Mini 44

Arduino Pro 46

Arduino Fio 47

Shields 49

Arduino GSM Shield 49

Arduino Ethernet Shield 50

Arduino Wi-Fi Shield 51

Arduino Wireless SD Shield 52

Arduino Motor Shield R3 53

Arduino Wireless Proto Shield 54

Arduino Proto Shield 55

Kits 57

Starter's Kit 57

Accessories 61

Arduino TFT LCD Screen 61

USB Serial Light Adapter 63

Mini-USB Adapter	64
Arduino Software	66
Getting Started	**67**
Installation Process	68
Windows OS	68
Mac OS	69
Linux OS	70
Arduino Development Environment	71
Sketches	71
Sketchbook	72
Tabs	73
Library	73
Language	73
Installing Additional Arduino Libraries	75
Automatic Installation	75
Manual Installation	77
Troubleshooting Problems	79
Creating Your Own Arduino	**82**
Breadboard	83
Developing Sketches	85

>   Other Important Information 88
> Some Examples 90
> >   Periodic Blinky LEDs 91
> > >   Assembling the Hardware 91
> > >   Writing and Uploading the Sketch to the Arduino 92
> >   Non-Periodic Blinky LEDs 95
> > >   Assembling the Hardware 95
> > >   Developing the Sketch 95
> Buying Arduino Products 98
> >   Arduino Store 99
> >   Distributors 101
> Conclusion 103

Introduction to Arduino

This is the era of rapid technological advancements which opens the doorway of opportunity for infinite possibilities. It is the computer age and is responsible for the advancements made in science and living. Previously, computers used to be enormous, gigantic pieces of machinery that took entire rooms to store. Today, you can fit them in the palm of your hands. Similarly, about a decade ago, the concept of smart phones was a shady dream. Today, it is a most-sought after common gadget of daily use.

The mechanization processes have made the lives of people substantially easier. The laborious hours have been transformed into effortless feats. The hefty machineries have been changed into convenient, small-sized and efficient packages of helping tools. It will not be wrong to say that the modern man can simply not survive in the non-technological era.

While talking about mechanization, it will be unjust not to mention automation. It is now possible to automate almost every activity. The first uses of automation were seen in the mass production lines where the pieces were fit together by machines. Today, it is possible to develop and administer similar protocols for everyday chores. What is more, you can develop your own protocols as well. This is where Arduino steps in.

Arduino is basically a single board microcontroller which comes fitted with the necessary circuitry, microprocessor, I/O circuits, clock generator, RAM and a stored program memory. This means once you have developed your personalized application program, you do not need to toil relentlessly on building the hardware controller. Arduino takes care of the hardware and ensures you are up and functional with your protocol in the minimum time.

For the intelligent and technical mind, this is a promising product which can create wonders. It can truly allow you to express your creativity and transform figments of your imagination into reality. You can use the Arduino products to create your own light, sound and temperature related circuits to suit different purposes in and around your residence and office.

Contained in this booklet is some information about Arduino. It is a starter's guide and will help you develop basic electronic objects – so you can get the gist of it and develop the complex projects without external help.

What is Arduino?

The simplest of definitions read, "Arduino is a single board microcontroller". In the more technical language, it is called as "an open-source electronics prototyping platform". Nevertheless, the product remains unchanged.

To begin with, Arduino is a circuit board which comes with a pre-mounted processor (as the CPU in computers but in a minute form), memory (as in RAM) and the input/output peripherals. The input/output peripherals allow you to sense and control the external environment.

So what is an Arduino? It is a blue colored board resembling the picture below.

It can be easier understood as a variation of the traditional computers – just smaller in size. The formulation of this board is such that it can sense inputs from the external environment and produce predetermined outputs that it is programmed to perform. If you are familiar with the term "physical computing", you can most easily relate it with an Arduino.

The Arduino is capable of creating the relationship between the world of electronics and the physical tangible world in a literal manner – it serves as a gateway between the two. It can be connected with all kinds of sensors, LEDs, sound systems, temperature regulators and motors directly.

Moreover, the memory on the Arduino board is capable of holding the programs you can easily develop on your computers. The Arduino board can be connected with your computer system via the universal serial bus (USB) port. It can be seen towards the left side of the Arduino image as a black colored box mounted on the board. This allows you to transfer the information from your computer onto the Arduino easily.

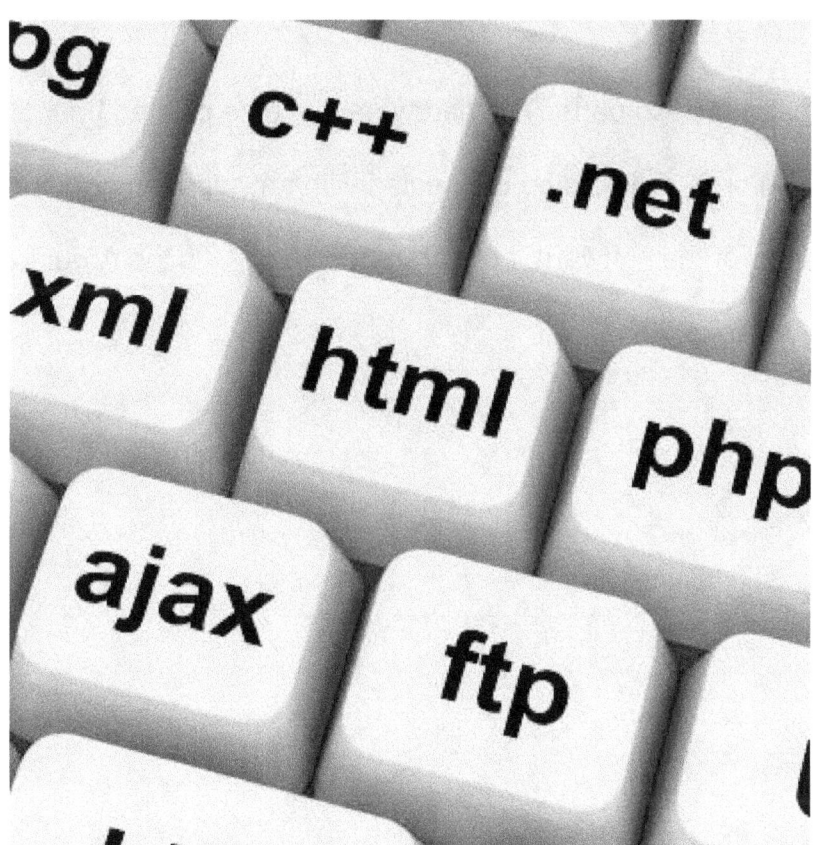

The programs can be developed using any programming languages that you have a command on. The most popularly used programming languages are Java and C/C++ though you can experiment with other languages as well. You can download the Arduino interface of writing the programs for free from their website. Once the programs have been written, they can be transferred and stored on your Arduino board easily.

Your Arduino board is capable of operating in collaboration with your computer or as a stand-alone unit – it only needs power to function, not a computer screen or keyboard or any other interface of any sort.

It is important to note that the brief size and capacity of Arduino is not built to handle vast amounts of electrical current and voltage. It can sense and manage digital and analogue voltages between 0 and 5 volts. It is therefore advised not to use Arduino boards in a heavy-duty setting.

To sum it up, an Arduino basically has two parts. The hardware is a blue colored board with all necessary chips mounted on it, ready to be connected with any input and output devices. The second part consists of the software – the program which tells the Arduino what to do, when to do it and how to do it.

Uses of Arduino

The possibilities with Arduino are infinite. It is not possible to restrict the uses of Arduino to a specific number of options. It depends on your creativity and innovation how you can use the Arduino to cater to a specific need.

The Arduino is an intermediary – a tool to meet greater ends. With a little technical know-how, you can even use the Arduino to prepare your evening cup of tea precisely at the moment when you enter the house!

Your Arduino is the present of the robotic world. The level of automation that you can achieve with this mini-robot knows no boundaries. With Arduino, it is even possible to transform the science fiction into practical reality.

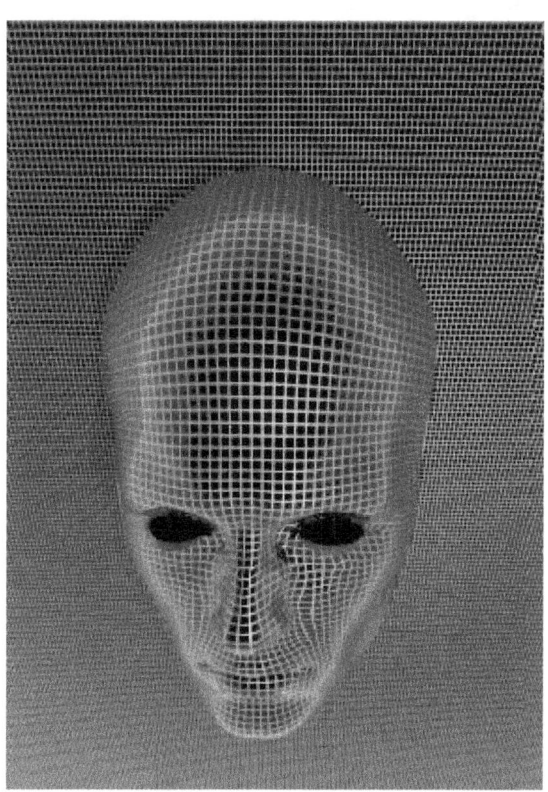

The Arduino is designed for use by students, artists, designers and hobbyists – anyone with a little technical knowledge who would like to experiment and create a few unique electronic creations.

Here are a few uses of Arduino to highlight its possibilities:

1. Sound Amplifiers
2. Pulse Sensor
3. 360 degree display projection
4. LED integrated shoulder pads, dresses and other clothing materials.
5. Fancy light shows
6. Musical Instruments' Tuners
7. Weather controlled lamps
8. Self-balancing objects
9. Playful mirrors
10. Time delay
11. Small robots
12. Prosthetic body parts
13. Automated drivers for cars
14. Flight simulators
15. Blinking displays
16. Potentiometers
17. Thermostats
18. Visual impairment aid
19. Remote controlled objects

20. Fully fledged robots

And lots more!

The possibilities of an Arduino are still under discovery phase. It all depends on you – what would you like the Arduino to do for you?

NOTE: The Arduino used to create each electronic item may be different and hence adapted to handle the power and voltages required to operate complex programming protocols.

Arduino Products

There are a number of Arduino products available in order to cater to the different requirements of each usage. All the products are unique and electronically different from one another. Also, some products are good for the beginners' experimentation while some can be used for the advanced level projects. The products are widely divided into four categories – boards, shields, kits and accessories. This section covers the technical aspects of each product and the various differences between them.

Boards

This category simply covers the microcontrollers. There are currently 19 products available in this category.

Arduino Uno

There are four variations of Arduino Uno:

1. Arduino Uno

Arduino Uno – Front

2. Arduino Uno SMD

Arduino Uno SMD – Front

3. Arduino Uno R2

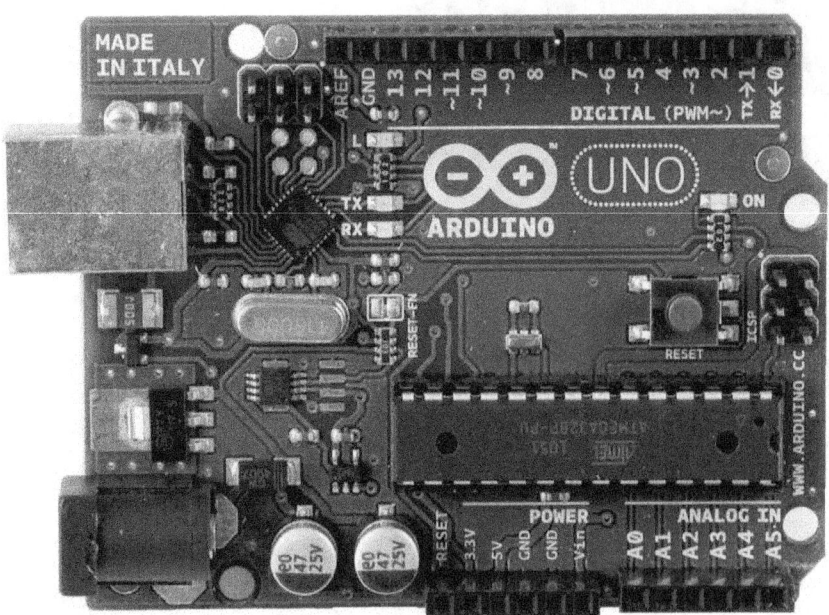

Arduino Uno R2 – Front

4. Arduino Uno R3

Arduino Uno R3 - Front

You can observe the slight differences between these four varieties.

The specifications are as follows:

Microcontroller: ATmega328

Operating Voltage: 5V

Input Voltage (Recommended): 7-12V

Input Voltage (Limits): 6-20V

Digital I/O Pins: 14

Analogue I/O Pins: 6

DC Current per I/O Pin: 40mA

DC Current for 3.3V Pin: 50mA

Flash Memory: 32KB

SRAM: 2KB

EEPROM: 1KB

Clock Speed: 16MHz

Size: 2.7 * 2.1 inches

Other Features: 16MHz Ceramic Resonator, Regular USB Connection, Power Jack, ICSP Header and Reset Button.

Arduino Leonardo

Arduino Leonardo - Front

The specifications are as follows:

Microcontroller: ATmega32u4

Operating Voltage: 5V

Input Voltage (Recommended): 7-12V

Input Voltage (Limits): 6-20V

Digital I/O Pins: 20

PWN Channels: 7

Input Analogue Channels: 12

DC Current per I/O Pin: 40mA

DC Current for 3.3V Pin: 50mA

Flash Memory: 32KB

SRAM: 2.5KB

EEPROM: 1KB

Clock Speed: 16MHz

Size: 2.7 * 2.1 inches

Other Features: 16MHz Crystal Oscillator, Micro USB Connection, Power Jack, ICSP Header and Reset Button.

Arduino Due

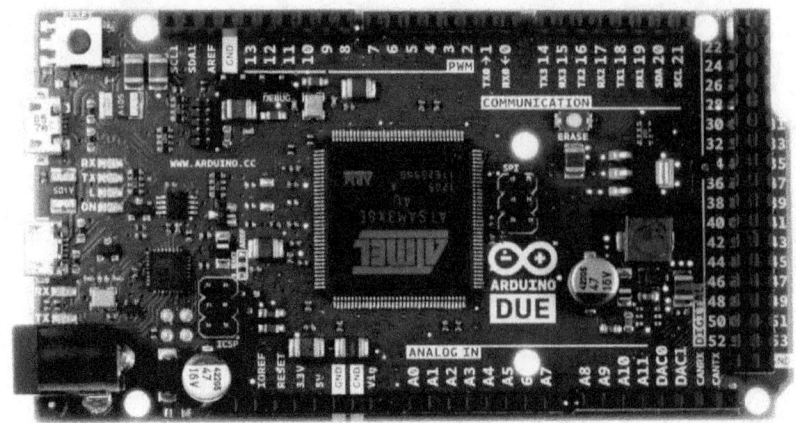

Arduino Due – Front

The specifications are as follows:

Microcontroller: AT91SAM3X8E

Operating Voltage: 3.3V

Input Voltage (Recommended): 7-12V

Input Voltage (Limits): 6-16V

Digital I/O Pins: 54

Analogue Input Pins: 12

Analogue Output Pins: 2

Total DC Output Current on I/O Pins: 130mA

DC Current for 3.3V Pin: 800mA

DC Current for 5V Pin: 800mA

Flash Memory: 512KB

SRAM: 96KB

Clock Speed: 84MHz

Size: 4.0 * 2.1 inches

Other Features: 4UARTs, USB OTG Capable Connection, 2DAC, 2TWI, Power Jack, SPI Header, JTAG Header, Reset Button and Erase Button.

Arduino Yun

Arduino Yun – Front

The specifications are as follows:

Microcontroller: ATmega32u4

Operating Voltage: 5V

Input Voltage: 5V

Digital I/O Pins: 20

PWM Channels: 7

Input Analogue: Channels: 12

DC Current per I/O Pin: 40mA

DC Current for 3.3V Pin: 50mA

Flash Memory: 32KB

SRAM: 2.5KB

EEPROM: 1KB

Clock Speed: 16MHz

Size: 2.7 * 2.1 inches

Other Features: Ethernet, Wi-Fi, micro USB Connection, micro SD card slot, 16MHz crystal oscillator, ICSP Header, and 3 Reset Buttons.

Arduino Micro

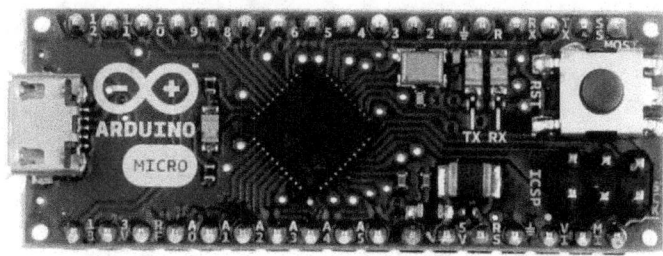

Arduino Micro – Front

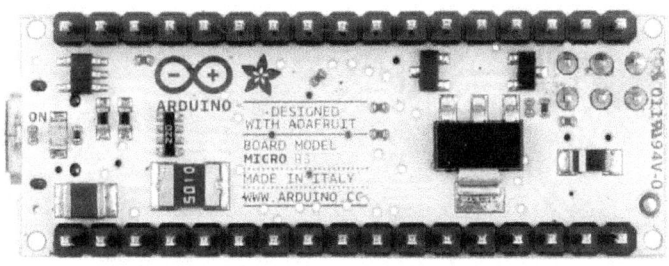

Arduino Micro – Back

The specifications are as follows:

Microcontroller: ATmega32u4

Operating Voltage: 5V

Input Voltage (Recommended): 7-12V

Input Voltage (Limits): 6-20V

Digital I/O Pins: 20

PWN Channels: 7

Input Analogue Channels: 12

DC Current per I/O Pin: 40mA

DC Current for 3.3V Pin: 50mA

Flash Memory: 32KB

SRAM: 2.5KB

EEPROM: 1KB

Clock Speed: 16MHz

Size: 4.8 * 1.7 centimeters

Other Features: 16MHz Crystal Oscillator, Micro USB Connection, Power Jack, ICSP Header and Reset Button.

Arduino Robot

Arduino Robot – Top View

The specifications are as follows:

Microcontroller: ATmega32u4

Operating Voltage: 5V

Input Voltage: 5V

Digital I/O Pins: 5

PWN Channels: 6

Input Analogue Channels: 4

Input Analogue Channels (Multiplexed): 8

DC Current per I/O Pin: 40mA

Flash Memory: 32KB

SRAM: 2.5KB

EEPROM (internal): 1KB

EEPROM (external): 512bit

Clock Speed: 16MHz

Size*:* 19cm in diameter

Other Features: USB Connection, Keypad, Knob, Full Color LCD, SD Card Reader, Speaker, Digital Compass, Soldering Posts and Prototyping Areas.

Arduino Esplora

Arduino Esplora – Front

The specifications are as follows:

Microcontroller: ATmega32u4

Operating Voltage: 5V

Flash Memory: 32KB

SRAM: 2.5KB

EEPROM: 1KB

Clock Speed: 16MHz

Size: 6.5 * 2.4 inches

Other Features: 16MHz Crystal Oscillator, Micro USB Connection, Reset Buttons, and Four Status LEDs.

Arduino ADK

There are two versions of the Arduino ADK.

1. Arduino ADK

Arduino ADK – Front

2. Arduino ADK R3

Arduino ADK R3 – Front

The specifications are as follows:

Microcontroller: ATmega2650

Operating Voltage: 5V

Input Voltage (recommended): 7-12V

Input Voltage (limits): 6-20V

Digital I/O Pins: 54

Input Analogue Pins: 16

DC Current per I/O Pin: 40mA

DC Current for 3.3V Pin: 50mA

Flash Memory: 256KB

SRAM: 8KB

EEPROM: 4KB

Clock Speed: 16MHz

**Size*:* 4.0 * 2.1 inches

Other Features: 4UARTs, 16MHz crystal oscillator, USB Connection, Power Jack, ICSP Header and Reset Button.

Arduino Ethernet

Arduino Ethernet – Front

The specifications are as follows:

Microcontroller: ATmega328

Operating Voltage: 5V

Input Voltage Plug (Recommended): 7-12V

Input Voltage Plug (Limits): 6-20V

Input Voltage POE (Limits): 36-57V

Digital I/O Pins: 14

Analogue Input Pins: 6

DC Current per I/O Pin: 40mA

DC Current for 3.3V Pin: 50mA

Flash Memory: 32KB

SRAM: 2KB

EEPROM: 1KB

Clock Speed: 16MHz

Size: 2.7 * 2.1 inches

Other Features: 16MHz Crystal Oscillator, RJ45 Connection, Power Jack, ICSP Header and Reset Button.

Arduino Mega 2560

There are two versions.

1. Arduino Mega 2560

Arduino Mega 2560 - Front

2. Arduino Mega 2560 R3

Arduino Mega 2560 R3 - Front

The specifications are as follows:

Microcontroller: ATmega2560

Operating Voltage: 5V

Input Voltage (Recommended): 7-12V

Input Voltage (Limits): 6-20V

Digital I/O Pins: 54

Analogue I/O Pins: 16

DC Current per I/O Pin: 40mA

DC Current for 3.3V Pin: 50mA

Flash Memory: 256KB

SRAM: 8KB

EEPROM: 4KB

Clock Speed: 16MHz

Size: 4.0 * 2.1 inches

Other Features: 16MHz Crystal Oscillator, Regular USB Connection, Power Jack, ICSP Header and Reset Button.

Arduino Mini

Arduino Mini – Front

The specifications are as follows:

Microcontroller: ATmega328

Operating Voltage: 5V

Input Voltage: 7-9V

Digital I/O Pins: 14

Analogue I/O Pins: 8

DC Current per I/O Pin: 40mA

Flash Memory: 32KB

SRAM: 2KB

EEPROM: 1KB

Clock Speed: 16MHz

Other Features: 16MHz Crystal Oscillator.

LilyPad Arduino USB

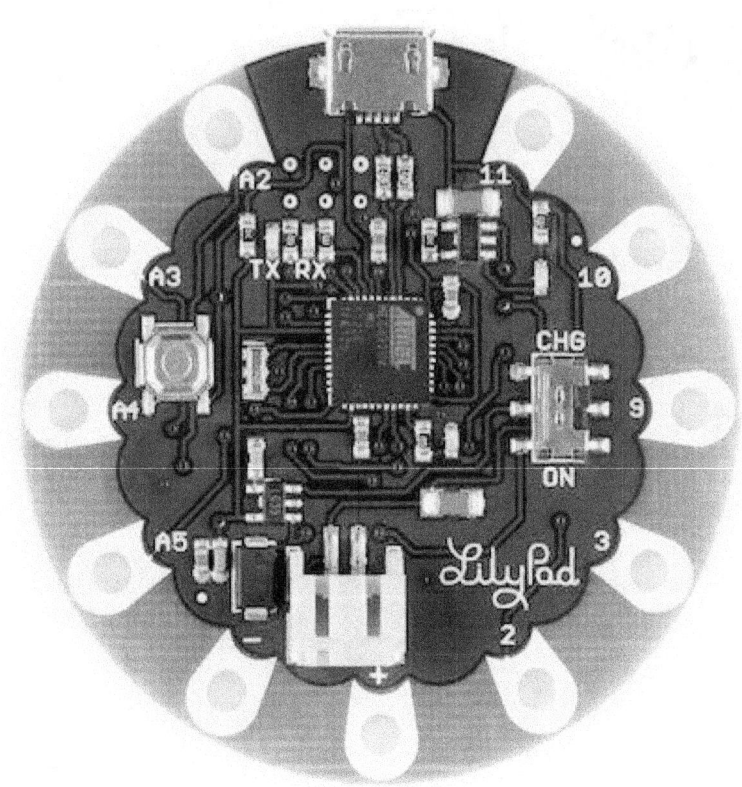

LilyPad Arduino USB - Front

The specifications are as follows:

Microcontroller: ATmega32u4

Operating Voltage: 3.3V

Input Voltage: 3.8-5.0V

Digital I/O Pins: 9

PWN Channels: 4

Analogue I/O Pins: 4

DC Current per I/O Pin: 40mA

Flash Memory: 32KB

SRAM: 2.5KB

EEPROM: 1KB

Clock Speed: 8MHz

Size: 2 inches in diameter

Other Features: 8MHz Resonator, Micro USB Connection, and Reset Button.

LilyPad Arduino Simple

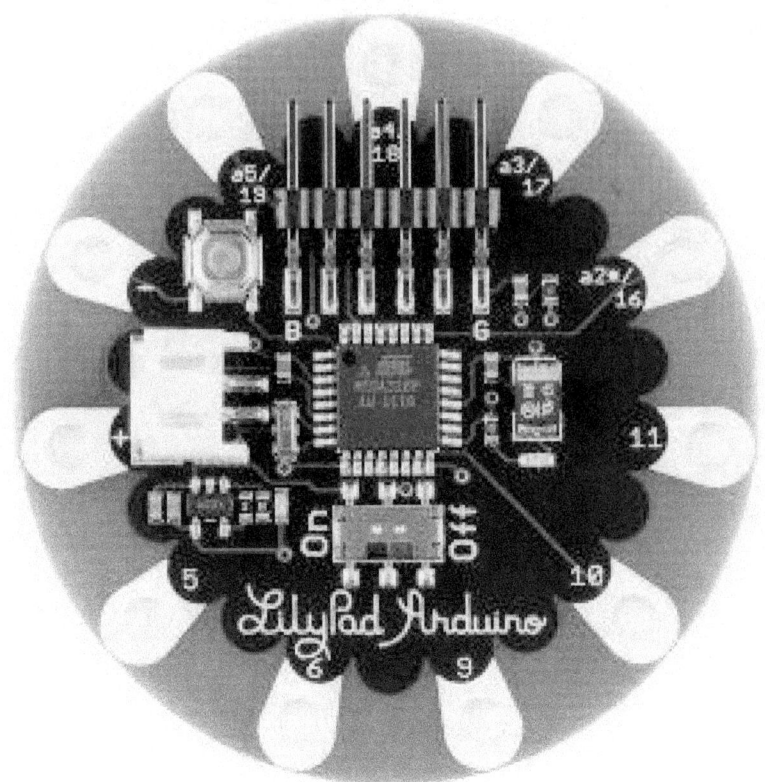

LilyPad Arduino Simple – Front

The specifications are as follows:

Microcontroller: ATmega328

Operating Voltage: 2.7-5.5V

Input Voltage: 2.7-5.5V

Digital I/O Pins: 9

Analogue I/O Pins: 4

DC Current per I/O Pin: 40mA

Flash Memory: 32KB

SRAM: 1KB

EEPROM: 1KB

Clock Speed: 8MHz

Size*: 2 inches in diameter

Other Features: JST Connector.

LilyPad Arduino SimpleSnap

LilyPad Arduino SimpleSnap - Front

The specifications are as follows:

Microcontroller: ATmega328

Operating Voltage: 2.7-5.5V

Input Voltage: 2.7-5.5V

Digital I/O Pins: 9

Analogue I/O Pins: 4

DC Current per I/O Pin: 40mA

Flash Memory: 32KB

SRAM: 1KB

EEPROM: 1KB

Clock Speed: 8MHz

Size: 2 inches in diameter

Other Features: Lithium Polymer Battery.

LilyPad Arduino

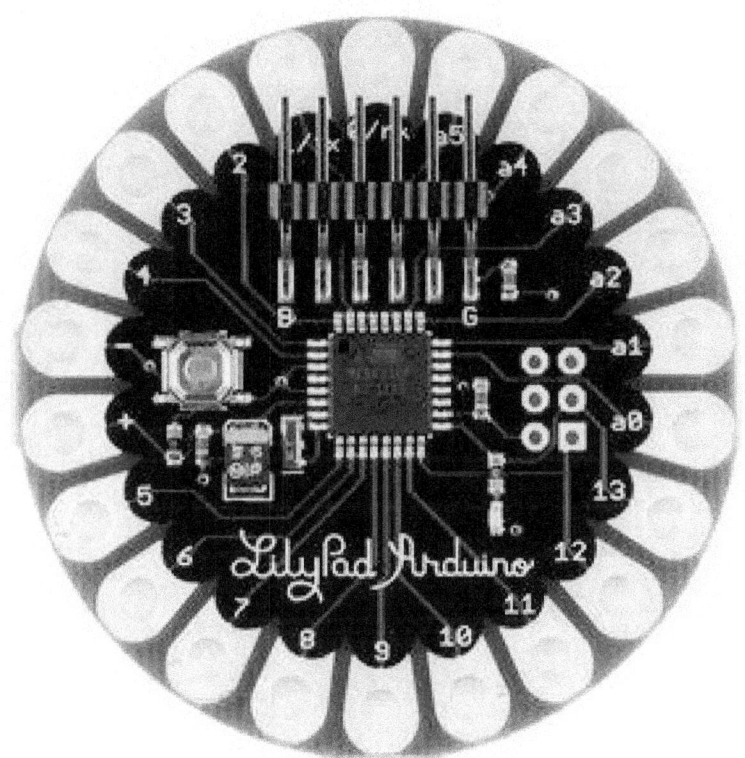

LilyPad Arduino – Front

The specifications are as follows:

Microcontroller: ATmega328V or ATmega168V

Operating Voltage: 2.7-5.5V

Input Voltage: 2.7-5.5V

Digital I/O Pins: 14

Analogue I/O Pins: 6

DC Current per I/O Pin: 40mA

Flash Memory: 16KB

SRAM: 1KB

EEPROM: 512 bytes

Clock Speed: 8MHz

Size: 2 inches in diameter

Arduino Nano

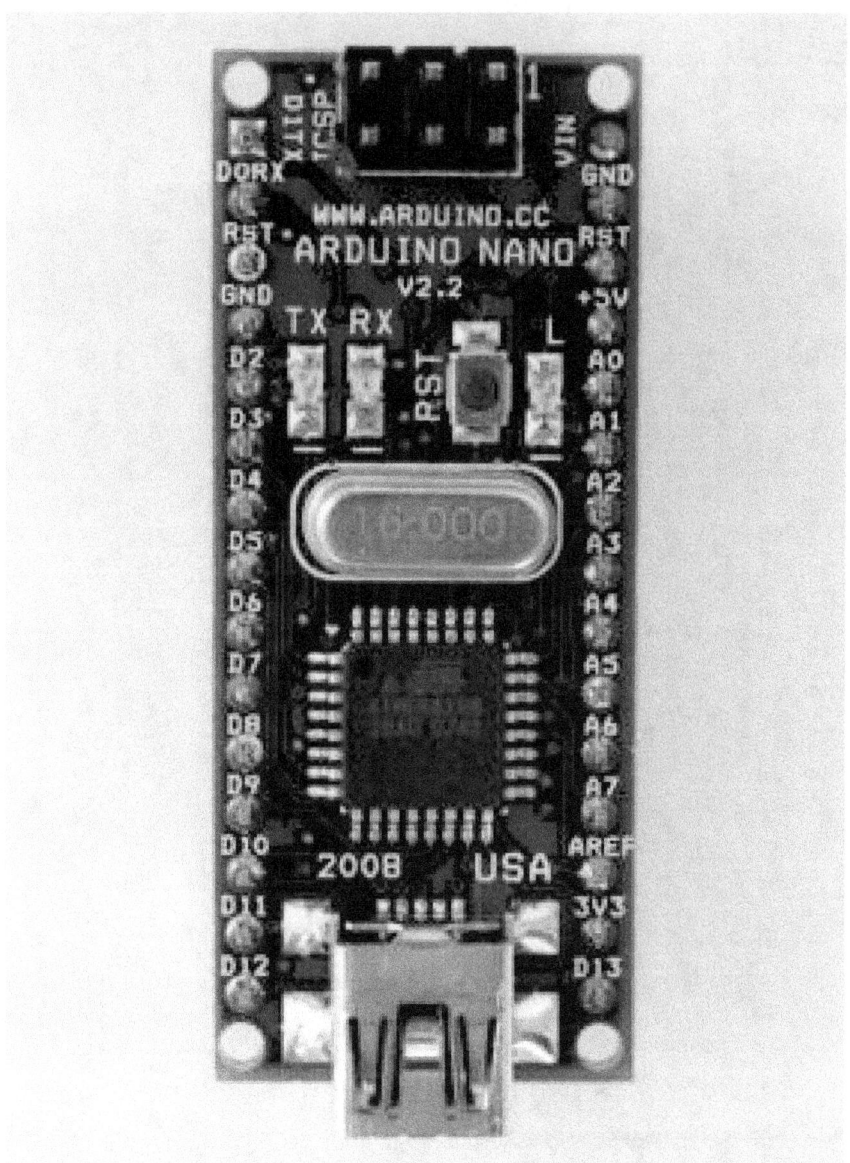

Arduino Nano – Front

The specifications are as follows:

Microcontroller: Atmel ATmega328 or ATmega168

Operating Voltage: 5V

Input Voltage (Recommended): 7-12V

Input Voltage (Limits): 6-20V

Digital I/O Pins: 14

Analogue I/O Pins: 6

DC Current per I/O Pin: 40mA

Flash Memory: 16KB

SRAM: 1KB

EEPROM: 512 bytes

Clock Speed: 16MHz

Size: 0.73 * 1.70 inches

Arduino Pro Mini

Arduino Pro Mini – Front

The specifications are as follows:

Microcontroller: ATmega168

Operating Voltage: 3.3 or 5V

Input Voltage: 3.35-12V

Digital I/O Pins: 14

Analogue I/O Pins: 8

DC Current per I/O Pin: 40mA

Flash Memory: 16KB

SRAM: 1KB

EEPROM: 512 bytes

Clock Speed: 8MHz

Size: 0.70 * 1.30 inches

Arduino Pro

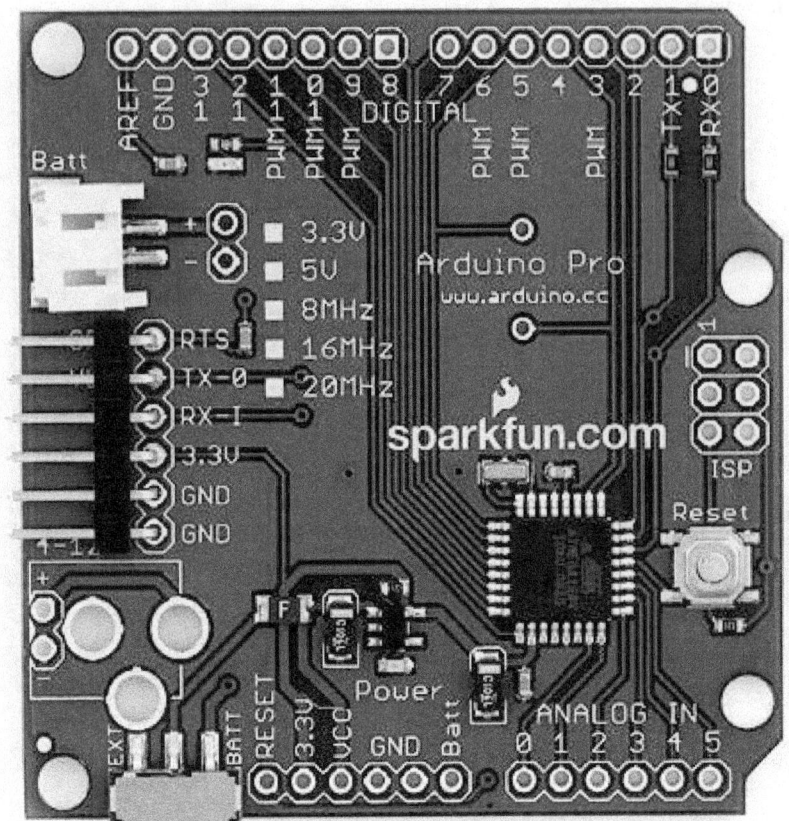

Arduino Pro – Front

The specifications are as follows:

Microcontroller: ATmega168 or ATmega328

Operating Voltage: 3.3 or 5V

Input Voltage: 3.35-12V

Digital I/O Pins: 14

Analogue I/O Pins: 6

DC Current per I/O Pin: 40mA

Flash Memory: 16KB

SRAM: 1KB

EEPROM: 512 bytes

Clock Speed: 8MHz

Size: 2.05 * 2.10 inches

Arduino Fio

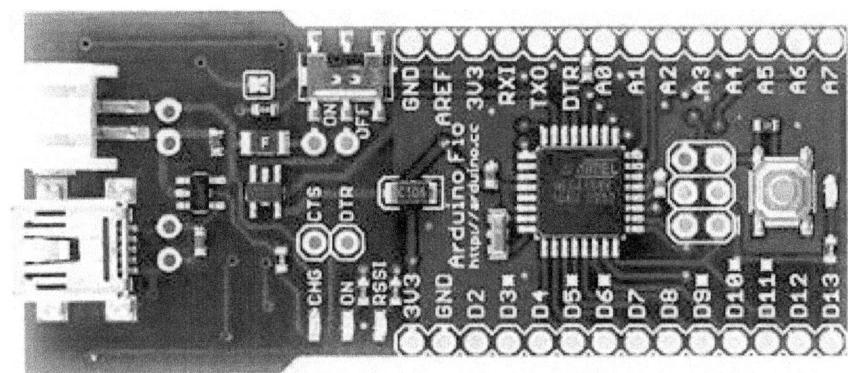

Arduino Fio – Front

The specifications are as follows:

Microcontroller: ATmega328P

Operating Voltage: 3.3V

Input Voltage: 3.35-12V

Input Voltage for charge: 3.7-7V

Digital I/O Pins: 14

Analogue I/O Pins: 8

DC Current per I/O Pin: 40mA

Flash Memory: 32KB

SRAM: 2KB

EEPROM: 1KB

Clock Speed: 8MHz

Size*: 1.1 * 2.60 inches

Shields

Shields come in handy when you want to connect your Arduino circuit board through the internet.

Arduino currently provides seven types of shields to serve different purposes.

Arduino GSM Shield

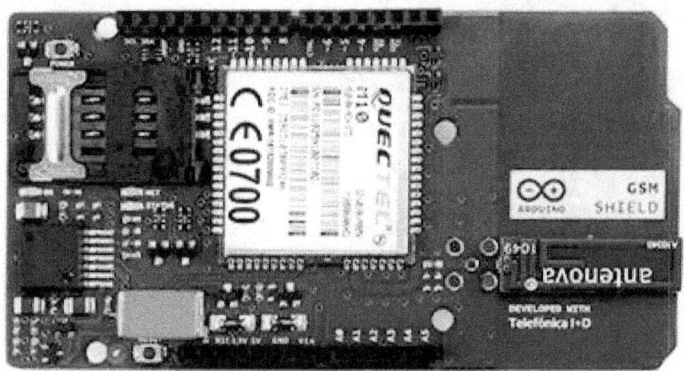

Arduino GSM Shield – Front

It connects your Arduino to the Internet using GPRS wireless network.

It requires:

1. An Arduino board
2. Operating voltage of 5V
3. It can be connected to the Arduino Uno.

The board can be externally powered with a 700-1000mA power supply. Powering the Arduino GSM Shield from the USB connection is not advised though it is possible.

There are three on board indicators – On, Status and Net.

Arduino Ethernet Shield

Arduino Ethernet Shield – Front

It requires:

1. An Arduino Board
2. Operating Voltage of 5V
3. Ethernet Controller W5100
4. Connection Speed of 10/100MB
5. Connection with Arduino on SPI port

The shield includes a reset controller to ensure the Ethernet controller is properly reset when power is switched on. It provides a standard RJ45 jack.

There are LED indicators on the Arduino Ethernet Shield to signify power on, presence of link, full duplex, 100M, Receiving data, Sending data, and network collisions.

Arduino Wi-Fi Shield

Arduino Wi-Fi Shield – Front

It requires:

1. An Arduino Board
2. Operating Voltage of 5V
3. Connection via 802.11b/g networks
4. WEP or WPA2 encryption
5. Connection with the Arduino on SPI port
6. On-board micro SD slot
7. ICSP headers
8. FTDI connection
9. Mini-USB connection for updating Wi-Fi Shield

The reset button refreshes both HDG104 and the Arduino board.

There are four informational LED indicators – L9, Link, Error and Data (being transmitted or received)

Arduino Wireless SD Shield

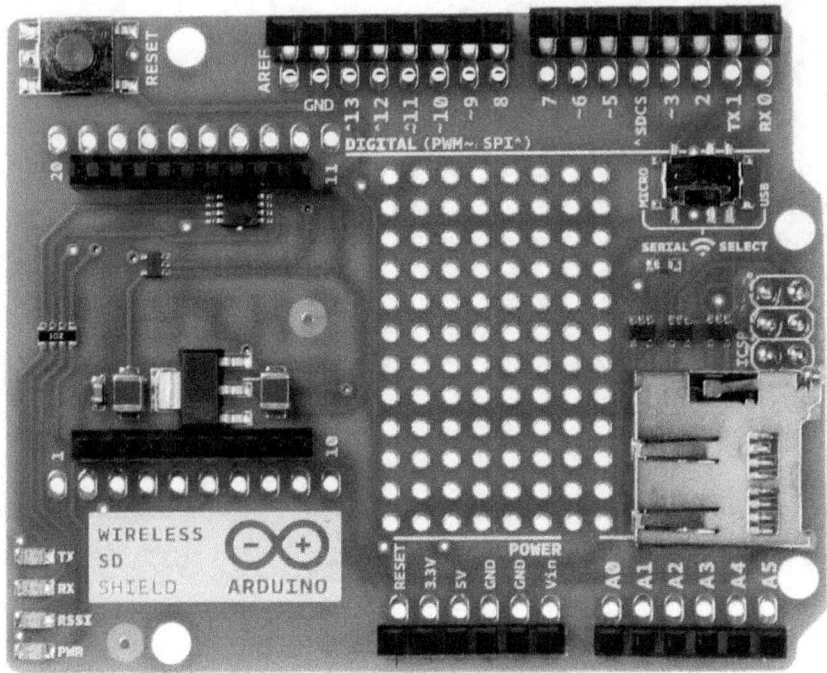

Arduino Wireless SD Shield – Front

It is exceptionally strong equipment which can be connected at a distance of 100-300 feet.

An SD card slot is available on the Arduino Wireless SD Shield which can be used to store information.

An on-board switch allows communication between the USB to serial convertor and the microcontroller.

The Arduino Wireless SD Shield uses the modules which have been adapted from XBee Modules.

Arduino Motor Shield R3

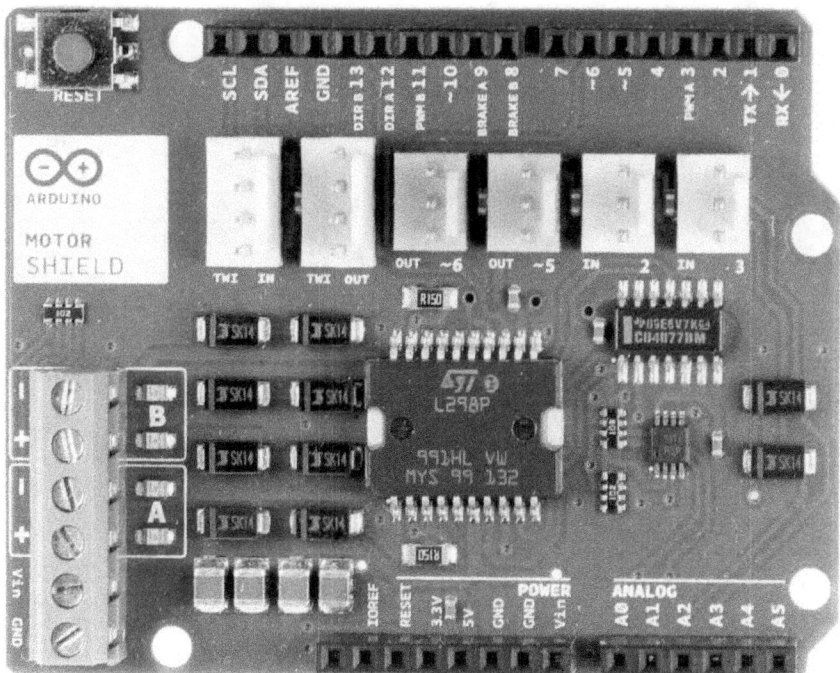

Arduino Motor Shield R3 – Front

This particular equipment has been built in such a way so as to serve as a dual full-bridge driver used to connect heavy duty motors, relays and solenoids to it.

You can connect two DC motors to this board simultaneously.

It requires:

1. 5-12V operating voltage
2. L598P – ability to connect to 2 DC motors or one stepping motor
3. A current of 2-4A from an external supply source
4. 1.65V/A current sensing
5. It also possesses a free running stop and brake function to stop the machinery immediately.

It is not advised to power the Arduino Motor Shield from a USB source.

Arduino Wireless Proto Shield

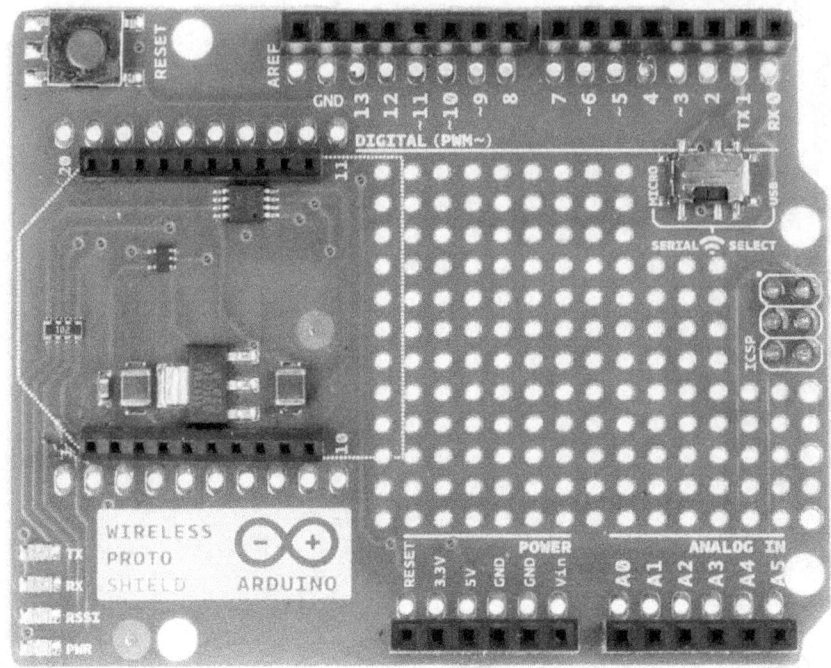

Arduino Wireless Proto Shield – Front

It allows your Arduino board to communicate wirelessly using a specific module.

It is also based on XBee modules from Digi. It can be connected from a distance of 100-300 feet while remaining within line of sight.

It does not have a SD socket. An on-board switch allows the connection to be made between the USB-to-serial convertor and microcontroller.

Arduino Proto Shield

Arduino Proto Shield – Front

It helps you to develop and customize circuits. It can be used with a breadboard in order to determine the efficacy of the circuit well before it is made permanent. It is important to note that the breadboard serves to make temporary connections only – this means if you would like to make more permanent connections, you might want to consider soldering or other such techniques.

It has some extra features like:

1. 1.0 Arduino Pinout
2. 1 Reset Button
3. 1 ICSP connector
4. 14 pin SMD footprint

5. 20 pin through hole footprint

The ICSP connector allows connections to be made with the SPI pins directly.

The dimensions of an Arduino Proto Shield are 2.7 * 2.1 inches.

Kits

Starter's Kit

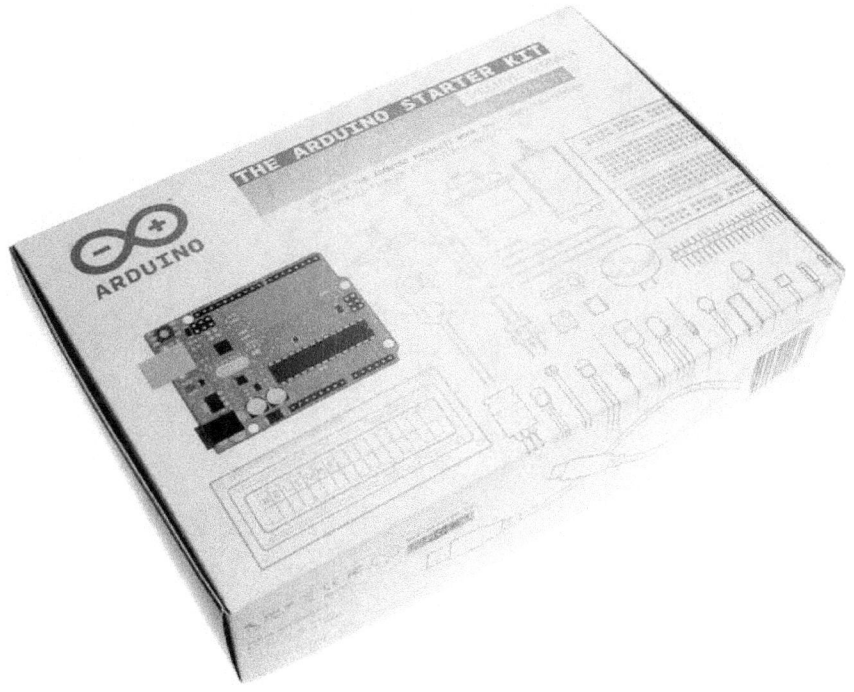

Arduino also offers a starter's kit for those beginners who would like to learn in their own way. It is an unparalleled way to learn Arduino while having hands-on experience with it.

The kit includes several of the most easily found electronic components which you can use to design creative projects. The kit also includes a book of 15 projects laid out in an easy-to-understand and easy-to-implement fashion. This basic knowledge will help greatly in teaching you how to use your own creativity with the Arduino circuits in order to make one-of-a-kind innovative products. It will also coach you about using sensors and actuators to control the physical world.

The book is divided into the following 15 parts:

1. Getting to know your tools: a basic know-how about each component and its usage. This will also aid in building the following projects in the book.
2. Spaceship Interface
3. Love-o-Meter
4. Color mixing Lamp
5. Mood Cue
6. Light Theremin
7. Keyboard Instrument
8. Digital Hourglass
9. Motorized Pinwheel
10. Zoetrope
11. Crystal Ball
12. Knock Lock
13. Touchy-Feel Lamp
14. Tweak the Arduino Logo
15. Hacking Buttons

The components included in the Starter's Kit are as listed under:

1. Arduino Projects Booklet
2. Arduino Uno Board
3. USB Cable
4. Breadboard
5. Wooden Base that is Easy-to-assemble
6. 9V Battery Snap

7. Solid Core Jumper Wires – 70 in quantity
8. Stranded Jumper Wires – 2 in quantity
9. Photo resistor
10. Potentiometer
11. Push Buttons
12. Temperature Sensor
13. Tilt Sensor
14. Alphanumeric LCD
15. LEDs (RGB, Red, Green, Yellow and Blue)
16. Small DC Motor
17. Small Servo Motor
18. Piezo Capsule
19. H-Bridge Motor Driver
20. Optocouplers
21. Transistors
22. Mosfet Transistors
23. Capacitors
24. Diodes
25. Transparent Gels
26. Resistors of different values

As it is evident, the equipments included in this Starter's Kit are extremely fundamental and will help you gain the basic know-how so that you can develop your own projects in the future. So if Arduino seems to be a horrific challenge to you, you can sit back and relax with the Starter's Kit and enjoy building wonderful projects with the guidance from projects booklet. In no time, you will observe how Arduino becomes easy-to-understand-and-manipulate to you.

Accessories

Arduino provides certain accessories to be used in combination with the Arduino boards so that their efficiency can be increased. As mentioned above, there are over 20 different Arduino boards, all with significantly different specifications. While some may be used without much add-ons, others might need one of the following accessories depending on the project you are trying to build.

Currently, there are three accessories.

Arduino TFT LCD Screen

Arduino TFT LCD Screen – Front

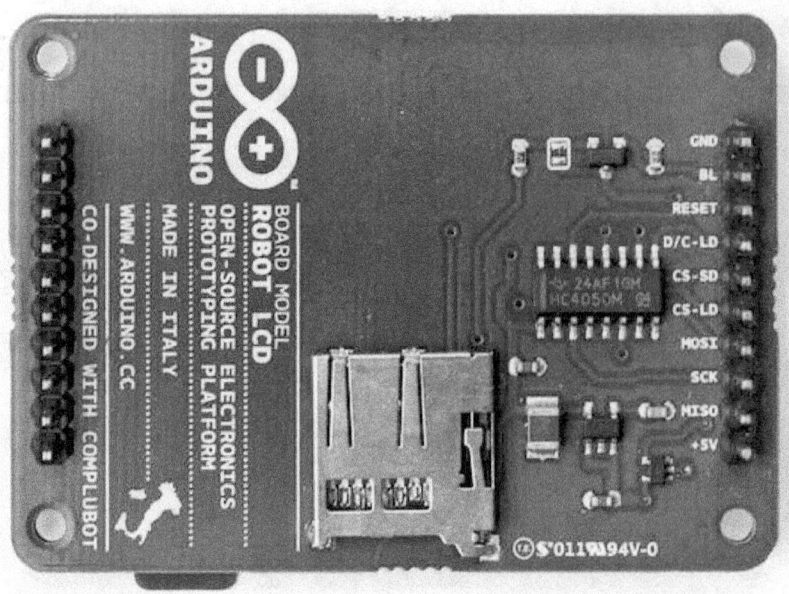

Arduino TFT LCD Screen – Back

The TFT Screen is a backlit LCD screen with headers. You can design text, images and shapes on the screen using the TFT Library.

The Arduino TFT LCD Screen comes with a built-in micro-SD Slot which can store the modules and images for display on the LCD screen.

It can be used with most of the Arduino boards including Esplora, Uno, Leonardo and Due.

The screen is 1.77 inches diagonally measured, with a resolution of 160x128 pixels. It requires a power of 5 VDC in order to operate this screen. The brightness of this LCD can be varied as per the requirements.

An extensive Library of TFT protocols is available with the Arduino TFT LCD Screen which can be loaded onto the device using the micro SD Slot.

USB Serial Light Adapter

USB Serial Light Adapter - Front

USB Serial Light Adapter – Back

You can connect the USB Serial Light Adapter directly into the Arduino Mini, Arduino Ethernet or other microcontrollers which will enable these boards to communicate with the computer systems. It requires a power of 5V in order to operate it.

The USB Serial Light Adapter possesses an on-board mini-USB connector with separate dedicated pins for receiving and sending data. The ground and reset buttons are also exposed on this device.

There are three status lights to indicate when the component is switched on, when it is receiving data and when it is sending it.

An on-board fuse limits the current flow to 500mA in order to prevent short circuits and damage.

The pinouts of the connector are compatible with a standard FTDI header.

Mini-USB Adapter

Mini-USB Adapter- Front

This component also serves the same purpose as mentioned above. It can be directly connected to the Arduino boards, allowing them to send and receive data from a host computer.

It requires a power source of 5V in order to operate it. It is simply the USB Serial Light Adapter reduced into a smaller size. All specifications are essentially the same.

Arduino Software

The Arduino software is available for free downloads on their website. It is an open-source software program which allows you to write your own codes and protocols, conditioning the circuits to perform accordingly.

It can be operated on Windows, Mac, OS X and Linux - the installation process may be different according to the operating system software. It allows you to write the software in a number of programming languages including Java, C/C++ and other languages.

There are 5 different softwares available on the Arduino website including:

1. Arduino 1.0.5
2. Arduino 1.5 Beta
3. Nightly Builds
4. Arduino IDE for Intel Galileo
5. Source Code

New softwares are continually developed and launched online, available for free downloading as and when they are ready for use. You can also access some previous versions on a separate link available on their "downloads" page.

With this, the Arduino products have all been introduced to you. Remember, the world of Arduino is full of possibilities – it only depends how you can conquer it!

Getting Started

Once you have been introduced to the wide array of Arduino products and the specific features of each, you are ready to proceed forward. This section will walk you through the introductory phase, the installation process and some of the other technicalities associated with Arduino that you need to be aware of in order to proceed with your wit and creativity.

At this moment, you are aware of the functioning of most Arduino products and you may have already ordered the starter's kit. Always remember that Arduino is full of possibilities – it is the present of a robotic future. If some parts of this book apparently fail to make sense, rest assured you will understand once you have indulged yourself into the world of Arduino.

Getting Started

Installation Process

The Arduino components cannot work according to your will until and unless you direct them. This is only possible with a program which is written and fed into your Arduino so that it knows what it has to do and when.

The production of the program requires you to download and install the Arduino interface/software (mentioned in the last section). The software can be downloaded for free from the official website of Arduino. According to your computer's operating system, follow the installation instructions in order to start writing your own programs for your special Arduino project.

Windows OS

For Windows, you will need an Arduino board and a USB cable.

1. Connect the board with your computer via the USB cable.
2. Download the relevant Arduino software from the website and install it prior to this connection.

3. Once you make this connection, you will be prompted to install the device drivers. Follow the installation wizard in order to ensure the Arduino understands your computer.
4. Now launch the Arduino application on your computer and write a program. You can also choose one from the existing library if you are doing it for the first time and do not know how to write the program.
5. Select your board and serial port to transfer the program to your Arduino.
6. And you are done! Keep experimenting to find out more about what your Arduino can and cannot do.

Mac OS

The procedure of installation with Mac OS is identical to that of Windows. Follow the same sequence of actions are written above and you should face no problems in running your Arduino board and program. If you face any problems, refer to the last part of this section for troubleshooting options.

Linux OS

Linux is comparatively complicated software. Installation of Arduino interface on Linux will require you to install other preliminary software prior to this. These preliminaries include "Java", "RXTX" and "avrdude". In addition to this, it is highly recommended for Linux users to install Arduino IDE 1.0.1 for its higher compatibility with Linux. The combination of this software will enable you to write your programs, access the existing programs on your Arduino and transfer newer ones on to it. It is important to develop this communication channel or your Arduino's relationship with your computer will not be a fruitful one.

Additional help is available on the Arduino website. You can also contact them via blog and chatting forums if you face any difficulty which you are not able to understand and overcome.

Arduino Development Environment

The Arduino Development Environment consists of a text editor which is used to write programs for Arduino, a message area, a text console, a toolbar with a list of commonly used functions and a series of menu bars for the less sought-after functions. This environment helps the Arduino hardware upload programs and communicate with the computers.

Sketches

```
/*
 Blink
 Turns on an LED on for one second,
 then off for one second, repeatedly.

 This example code is in the public domain.
*/

void setup() {
  // initialize the digital pin as an output.
  // Pin 13 has an LED connected on most Arduino boards:
  pinMode(13, OUTPUT);
}

void loop() {
  digitalWrite(13, HIGH);   // set the LED on
  delay(1000);              // wait for a second
  digitalWrite(13, LOW);    // set the LED off
  delay(1000);              // wait for a second
}
```

The picture above shows what a sketch looks like. It is just a sample.

The programs that you write are commonly known as 'sketches'. These are written in the text editor and are saved with an extension filename of ".ino". The text editor can receive text in any form.

The message area displays information regarding the written program. It can inform about the utility and functioning of the program. It can also display any errors observed by the environment in writing the program. The toolbars allow you to verify and upload programs to your Arduino, select the board and serial port, and also to save new Arduino files and open older ones for review.

In a nutshell, the Arduino Development Environment can perform the following actions:

1. Verify programs
2. Upload them to your Arduino board
3. Create new sketches
4. Open existing sketches for review and changes
5. Save new sketches and also the changes made to previous versions
6. Open the serial monitor

Sketchbook

The sketchbook is a place where all your sketches are stored. You can access your sketches from this directory, view them and change them at a future date if you so desire.

Tabs

You can open multiple files at the same time on your Arduino development environment. This will enable you to edit each of the files simultaneously and also to compile them if need be.

Library

The Arduino library provides extra functionality for use in sketches. All you need to do is include a specific part of the library into your program. So while uploading the program to your board, the specified portions of the library will also be shifted onto the Arduino board. This takes up space, and as you have seen already, there is severe memory limitations with each of the boards mentioned above.

Language

You can change the language of the Arduino development environment according to your needs. This will enable you to increase the efficacy of the programs for yourself.

Language Support

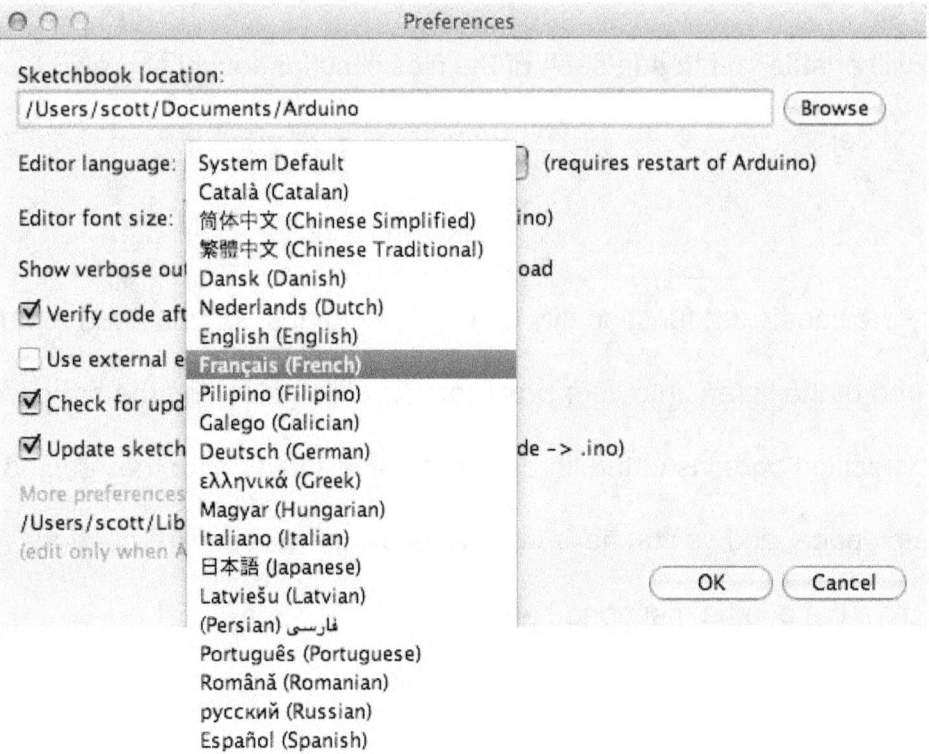

Moreover, once you have made a change, it will require you to restart your computer for the changes to become completely effective.

Installing Additional Arduino Libraries

Once you are through with the basic Arduino software interface and you are able to use the basic functions with proficiency, you can improvise further by the use of libraries.

Libraries are ready made protocols which you can use to further increase the functionality of your Arduino board. These codes help your Arduino board to communicate with external sensors like LCD displays.

The Arduino development environment comes with the basic functionalities pre-installed whereas additional libraries need to be downloaded from the website and installed on your computer for you to be able to use them in your Arduino programs.

There are hundreds of additional libraries available online which you can use to improvise your programs. Imagine the possibilities with such a vast library of resources.

Automatic Installation

The installation of additional libraries is possible if the files are available in .zip format. You can prompt your computer to initiate direct installation using the .zip file – you do not need to extract it.

Here is what you will need to do:

1. Select the "Sketch" tab from the menu bar.
2. A drop-down menu will be displayed. Select the option "Import Library".

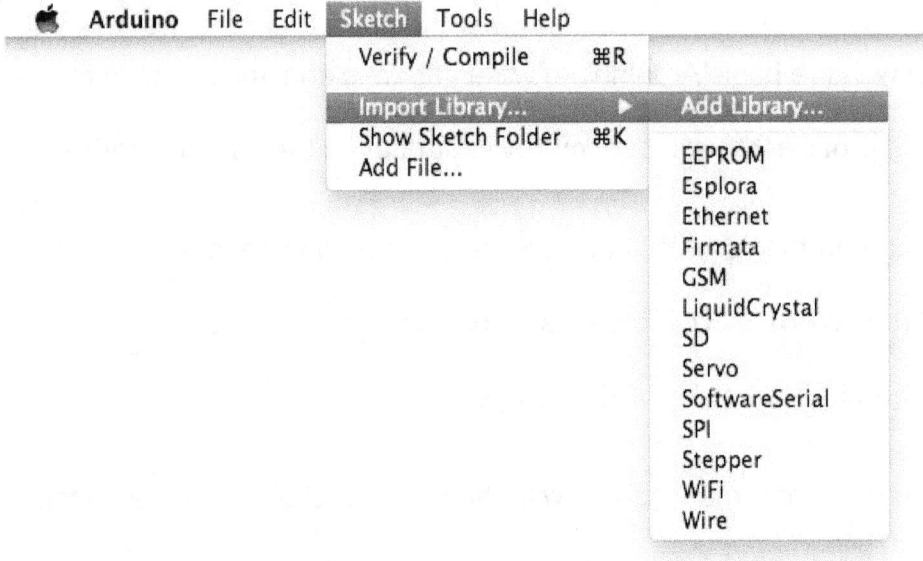

3. This will open an external dialogue box prompting you to select the destination where your file is stored. Select the file and click "Open".

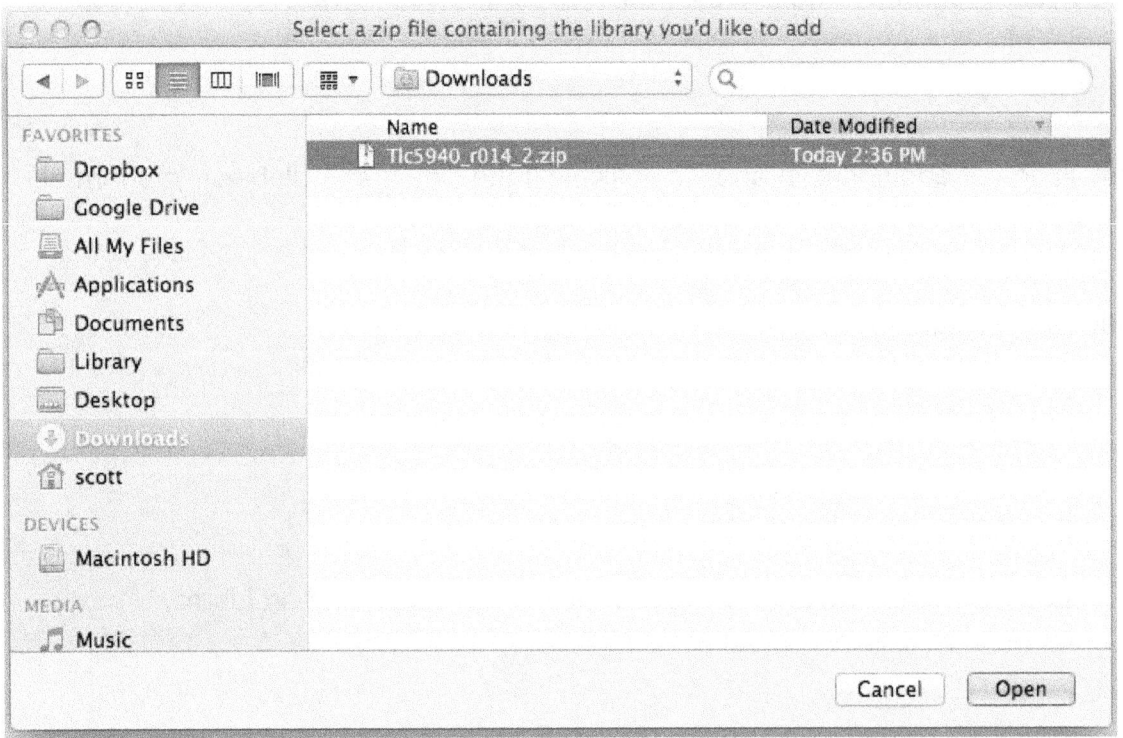

4. This step will take a little time. The software will automatically decode the .zip file and install the library for you to use in your programs.

5. Lastly, in order to access this newly installed library, you can access it by Sketch>Import Library>(Contributed Library). The Contributed library will be visible by the name of the library you have downloaded and installed.

6. The IDE will need to be restarted for the complete functionality of this software to be available.

If the automatic installation process does not work out for you, you can also opt to download and install it manually. This will require more steps; nevertheless, it will ensure the library is available and accessible to you.

Manual Installation

Manual installation of the Arduino library will need you to quit the application before you can proceed. The manual installation is not possible without performing this important action before any other.

1. Quit the Arduino application
2. Uncompress the .zip file. Inside this folder, you will see files with the extension names .cpp, .h and .txt. If these files are not available in the folder, you will need to create them so that the software can function properly.
3. Now you will need to copy these files into your Arduino libraries store. It will most likely be available in My Documents>Arduino>Libraries (in case of Windows). It will be a similar case for Mac OS and Linux as well.

4. If there are more files in the .zip folder, you should ensure you copy all of them onto this new location for maximum utility.
5. Now launch the Arduino application.
6. The newly installed library should be available in Sketch>Import Library section.
7. And you are done! You have successfully installed the additional library.

It is important to remember that you may need to install additional libraries only at an extremely advanced level. For beginners, the additional libraries can prove to be more of a hassle than anything else. Make sure you have a command on your basic Arduino skills before opting for this alternative. Once you are ready to experiment with your Arduino board's possibilities and you have the basic know-how, the additional libraries will be able to do wonders for you.

Troubleshooting Problems

It is possible for your Arduino program and software not to function as appropriately as you may want it to. This does not mean you need to panic about it. You can always try to troubleshoot these according to the long list of commonly faced problems listed on their website.

If the problems persist, you can contact the technical professionals on the Arduino blog and chat forums.

Here a few pointers you should keep in mind and make sure you have followed these to eliminate most of the problems faced.

1. Make sure all connections have been made properly. The Arduino board should have a power source (if it is not being powered by the computer), and the sending and receiving pins should be connected properly. If any of the connections are not made, the Arduino will not be able to communicate with the computer.
2. The IDE Arduino Development Environment should be correctly downloaded and installed on the computer. If you are using additional libraries, make sure they are installed and displayed in the Arduino Development Environment. If automatic installation does not work for you, you can try to install the libraries manually. If it still does not work for you, you should try downloading the library again. Feel free to contact the forum if the problem seems to persist.
3. Make sure you preliminary programs are compatible with the Arduino application. For instance, the latest versions of Java are built around 64 bit versions of native

libraries. However, most Arduino libraries work with the 32 bit versions of native libraries. Thus you need to ensure all software is compatible with one another to ensure the application and programming runs smoothly.

4. Some processes which are running in your computer will not allow the Arduino software to perform properly. Ensure that no such conflicting processes are running. For instance, with Windows, the process "LVPrcSrv.exe" does not allow the programs to upload to your Arduino board. In fact, it freezes your software. Make sure such conflicting processes are not running while you try to run your Arduino application.

5. Depending on your power supply, make sure your connections are made properly. The USB power option requires different settings on your Arduino board whereas if you like to power your board through external power, you will need to change the settings slightly. If your Arduino board is not powered properly, it will not be able to function.

6. If, for some reason (like improper shutdown), your Arduino.exe file does not run and instead displays an error, you can try running the run.bat file. This format may take some time to run but will nevertheless be accessible.

7. Your Arduino drivers need to be installed so that they are visible on your Arduino interface. If you have connected a particular board and it is not showing up on the Arduino Development Environment, it is probably because the drivers are not installed. Make sure you have installed the drivers properly when you connect the board.

8. If you are using any part of the built-in or additional libraries, make sure they are "included" in your program. This part of the library will be copied along with your program onto your Arduino board. If you have not included it, your program will not function properly.

9. Another important consideration with respect to libraries is the size of your program. It is very important for you to remember that the memory on your Arduino board is extremely limited. If your program is too large in size, it will not be transferred to your Arduino board. Consequently, your program will not work. Make sure you have removed any libraries that your program does not need to function.

10. If steady power supply is not available, it may cause the program to be wiped off from your Arduino board's memory. Make sure your Arduino board is powered through a steady power source.

11. If you seem to face a problem which has not been addressed in this list or in the troubleshooting list available on Arduino's official website, you can always discuss it with someone who has knowledge about Arduino. At the Arduino blogs, you can communicate with fellow Arduino users to find solutions to your problems. You can even contact the professionals at Arduino to help you resolve your problem. Your colleagues, teachers and supervisors are also a good source of troubleshooting problems. It is extremely rare to find a faulty Arduino board. If nothing at all seems to work, you should consider changing the board.

Hopefully, you will be able to build a lasting and fruitful relationship with your Arduino.

Creating Your Own Arduino

It has already been stressed innumerable times that the possibilities with Arduino are practically limitless. You can create your own Arduino boards with ease if you know what you want to make and how. There are hundreds of examples and basic recipes of building your Arduino projects. While these are a good source of getting hands-on experience, they also enable you to understand the role and functions of each component.

Creating Your Own Arduino

Breadboard

Before carrying forward to some examples, you need to familiarize with the breadboard.

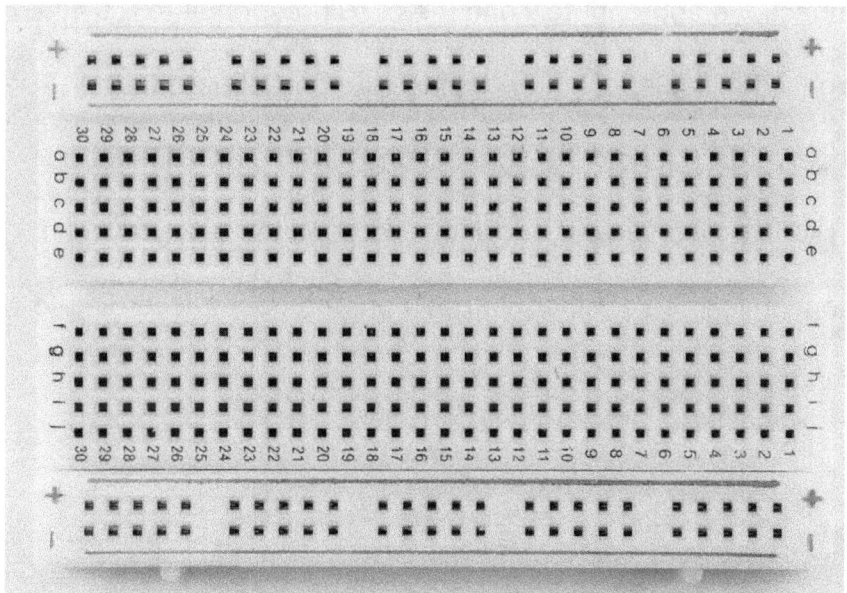

A breadboard resembles the one shown in the picture above. It is a board made out of fine durable plastic with several aligned square-shaped holes on its face. You can clearly see the letters and numbers written on the breadboard. These help identify the connections you make. The positive and negative charges signify the flow of current in the breadboard. You need to be careful about the positive and negative connections you make – if they are not in accordance with the general flow of current inside the breadboard, your circuit will not function.

The breadboard allows you to build temporary connections. It is best used for experimentation as it shows the outcome of your connections without making them permanent. In case a wrong connection has been made and you do not see the results you hoped for, you can easily correct them by changing the connections if you use a breadboard. Soldering and printing boards also perform more or less the same function.

However, changing connections after soldering is difficult and it is all the more impossible if you have printed the board.

Once you have built the connections appropriately and the circuit responds exactly the way you wanted it to, it is highly advisable for you to replace the breadboard with a permanent solution. Breadboards do not hold the connections for long and cannot be used robustly. If you would like to preserve your circuit connections the way you have developed them, your best option is to solder them or get a printed board according to the connections you have made.

Developing Sketches

There are few things you need to know about developing sketches. The programming language is quite different from our communication languages. The programming languages are governed by some protocols. If these protocols are not followed, you will not be able to achieve the desired results. It is therefore advisable to be familiar with these before you start with developing your own sketches.

Here are a few rules to get you started:

1. All sketches need to have two important parts – the setup and the loop. Hence the simplest of programs need to look like this:

    ```
    void setup()
    {
    }
    void loop()
    {
    }
    ```

    The code between the first set of curly brackets will be run the first time your Arduino board is plugged in and your program is run. It is similar to the installation of any software.

    The second part – loop – informs the Arduino what it has to perform repeatedly till its power supply is intact. For instance, you can instruct your Arduino to constantly blink an LED – and it will do so till the power supply is removed from your Arduino board. This is the actual program protocol which tells your Arduino what it has to do, how, when and in what manner.

2. Anything that you write in a single line after using double backslashes "//" will be ignored by the program. This is usually done to support human text. You can write about a specific function that you have entered – this will help in debugging your program later as well as informing any human reader about what your program is supposed to do. It is important to note that this function works only as long as your statements fit in one line. For longer texts, the next pointer can be used effectively.

3. If your helpful information exceeds the character limit of one line, you can use one backslash followed by a star to mark the beginning of your help text – "/*". Make sure you use the reverse signs to mark the end of your text or else the entire program will be ignored – "*/". Anything written in between /* and */ will not be processed by the processor – it is overlooked by the program decoder.

```
/* This is to
inform you how
you can instruct
the Arduino
processor to
ignore a portion
of your program
*/
```

4. The setup codes and loop functions need to be written inside curly brackets. The program will not be functional until and unless there is a code between the curly brackets. Moreover, the usage of some other form of brackets (parenthesis or square brackets) will not be able to serve the same purpose as curly brackets.

5. Every line in the program should end with a semicolon or the program will not run. This is often the mischievous reason behind the non-functioning of programs. However, semicolons are not required after the closing curly bracket.
6. All open parenthesis should be accompanied with a closed parenthesis – () – or the program will not function properly.

Other Important Information

There is more to making your own Arduino other than using a breadboard and learning how to develop your sketches. Here is some more helpful information you should keep in mind while planning to develop your own Arduino project.

1. There are some components required in order to make anything out of your Arduino board. The starter's kit has an adequate amount of each. If you do not want to purchase the starter's kit for any reason, you can purchase the components listed above in the section "Products>Kits>Starter's Kit" separately.

2. Your Arduino and breadboard will need a holder. Even though this is not compulsory, the holder can help you in keeping your Arduino board and breadboard fixed together so the circuit is not disturbed during preparation and otherwise. The wires are only placed lightly into the breadboard's holes and hence are not strong enough to withstand any pressure.

3. The best basic Arduino board to use is the Arduino Uno. It is equipped will all the necessary components including the USB connection and also has a dedicated portion for the installation for battery for standalone operations. While you can use any other board as effectively as Arduino Uno, it is generally advisable for beginners to initiate their Arduino adventure with this basic form of Arduino board.

4. Be very careful to turn off your power source before you change the circuit. Failure to do so can cause short circuits and may also damage your Arduino board and other components used to develop your projects. It can also cause

discomfort to you in the form of electrocutions by the minor current flowing through the board.

5. Make sure you read about the capacities of each electrical component that you use. For instance, your LEDs can bear a certain amount of current and voltage – if you power it with more, the LED will burn out and be damaged permanently. In case of other equipments, the concern is even more as they are likely to be more expensive and sensitive than an LED. Make sure you read about the specifications of each component before you use them in your projects.

Some Examples

The best way to learn Arduino is through examples. Grab your Arduino Starter's Kit now and start building the very first of your projects. This will not only be able to give you an adequate hands-on experience, it will also help you understand the working mechanism behind Arduino in general. So you can move forward with your own projects easily.

Here are a few simple circuits to get you started. Remember, there are two parts to building your Arduino project – assembling the hardware and programming the software. Each is incomplete without the other. It is therefore important for you to follow all the steps perfectly in order to develop a working Arduino project.

Periodic Blinky LEDs

The most basic Arduino program in our view is programming the LEDs to switch on and off progressively after timed intervals. All you need is the Arduino Uno board, some wires, an LED, a resistor (preferably 330 ohms), a breadboard and a holder. Here is what you need to do.

Assembling the Hardware

1. Attach your Arduino board and the breadboard to the holder using screws. This will ensure the circuit remains intact and in one place.
2. Use a wire to connect the Arduino board's pin GND (stands for ground) to the breadboard's right-most and bottom-most hole.
3. Use another piece of wire to connect Arduino board's pin titled 5V to the adjacent hole from the last connection. These two wires ensure the breadboard will be powered when the Arduino is plugged in.
4. Now connect the resistor to the breadboard. Place one leg of the resistor is column H and row 2 (generally called h2). The other leg needs to be connected in the column where Arduino's ground wire is connected (right-most column, same row).

   Since the resistor is not direction-sensitive, it can be connected in any way.
5. It is now time to connect the LED. Unlike the resistor, the LED is direction-sensitive and will only light up when connected in a specific direction. If you have performed all other actions perfectly and your LED refuses to light up, you can consider changing its connection 180 degree.

   Place one leg of the LED (shorter one) to column F row 2. This will be in the same row as the resistor.

Connect the longer leg of the LED in column F row 3.

6. Now use a piece of wire to connect H3 to the column where the 5V wire from the Arduino is connected (second last from right). Make sure the row is the same.

If you have performed the aforementioned steps appropriately, the LED should glow when you power up the Arduino board (without any program). This tells you that the LED is placed in the right direction. If this is not the case, consider changing the direction of the LED. This is done to check whether the connections have been made properly.

7. For your Arduino to be in control of this circuit, you will need to make a slight change. Previously, you attached the wire from H3 to the column where 5V wire from the Arduino is connected. Now change it – connect the wire from H3 to the pin labeled 13 on the Arduino board. This will put the Arduino in charge of the circuit. You are now ready to proceed towards developing the sketch.

Writing and Uploading the Sketch to the Arduino

1. You will need your computer with the Arduino development environment installed correctly on it. Refer to the previous sections to find out about how it is done.
2. Run the IDE Arduino development environment application. It will display a window with some space in the middle, toolbars and message boxes.

3. Write the following program in the text box.

```
1   const int kPinLed = 13
2
3   void setup ()
4   {
5           pinMode(kPinLed, OUTPUT);
6   }
7
8   void loop ()
9   {
10          digitalWrite(kPinLed, HIGH);
11          delay(1000);
12          digitalWrite(kPinLed, LOW);
13          delay(1000);
14  }
```

This is a 14 lines program. Remember, it is very important for your program to have the setup and loop parts, the codes in between the curly brackets and semicolons at the end of every line.

4. Be extremely careful in typing out the program. The language is extremely sensitive. Spelling mistakes and inappropriate capitalization will not allow the program to run smoothly. Spaces, tabs and blank lines do not have any impact on the program whatsoever.

5. Once you have written the program, the next step is to shift it onto your Arduino board.

Connect your Arduino board to your computer via the USB port. Make sure that the board drivers have been installed and the board is in sync with the computer. From the IDE Arduino development environment application, select the relevant board – Tools>Boards Menu.

Select your Arduino board from the menu and then click upload. The program will automatically be transferred onto your Arduino board.

When the program has been transferred completely, it will display the message "Done uploading". This means your board is programmed and ready to use.

6. Power up your Arduino board using the USB as power source or from the external battery connection. See your first project running smoothly.

In this "Blinky LED" project, you have connected one LED to the Arduino. You can further improvise this program by adding multiple LEDs and programming them to blink alternatively. In this way, you can develop your own light shows and blinky decoration pieces.

You can even attach the LEDs in the shape of words – like "Happy Birthday" – and time them to blink in several different ways. So you do not need to show the message, you can flash it!

Non-Periodic Blinky LEDs

It is not necessary for you to program your LEDs to blink at set intervals – you can program them to blink at different intervals each time. Make the LEDs blink faster, then slow down and repeat this cycle till the power source is removed. Here is what you will need to do.

Assembling the Hardware

The connections made in the case of periodic and non-periodic blinky LEDs is more or less the same. It is up to you how many LEDs you would like to connect.

Make sure you have checked the validity of the connections before you proceed towards attaching the Arduino and programming it. This will ensure your project runs smoothly.

Developing the Sketch

The development of sketch will be slightly different in this case. Follow the same steps as mentioned above except the program. Write the following program in the IDE Arduino development environment window:

```
1    const int kPinLed = 13;
2
3    void setup()
4    {
5            pinMode(kPinLed, OUTPUT);
6    }
7
8    int delayTime = 1000;
9
10   void loop()
11   {
12           delayTime = delayTime - 100;
13           if(delayTime <= 0)
14                   {
15                   delayTime = 1000;
16                   }
17   digitalWrite(kPinLed, HIGH);
18   delay(delayTime);
19   digitalWrite(kPinLed, LOW);
20   delay(delayTime);
21   }
```

This 21 lines program will instruct the LED to blink faster till the delay time between each blink has been reduced down to zero. At this time, the delay time will automatically be reset and the LED will be down to stage one.

If you use more than one LED, you can program each LED to follow a different pattern of time delay. This will enable you to build your own unique random light shows.

Food for Thought: if you feel like using LEDs to build your unique themed costume party dress, using this protocol will allow you to stand out amidst the crowd. Stay out of the ordinary with Arduino.

Another important pointer with respect to this program is that the same results can be achieved using different programming protocols. For example, you can use the commands ELSE, WHILE, TRUE/FALSE, and FOR instead of using IF. The programming statements will also change accordingly. This is just to emphasize that there are a hundred ways of achieving the same result when you are using an Arduino.

Buying Arduino Products

At this point, it is anticipated you would be wholly interested in getting your Arduino boards from somewhere to start building different Arduino projects immediately.

Such is the charm of Arduino – it makes you want to lay your hands on this wonderful piece of equipment and start building your own projects immediately. The possibilities with an Arduino are infinite.

You can instruct your Arduino to prepare your cup of tea while you are parking your car with just a message. Or you can train it to message you if your plants are too dry. You can truly let your imagination go wild with an Arduino.

There are two ways in which you can acquire your Arduino boards. Here is a little detail about each mode and how you can place your order.

Arduino Store

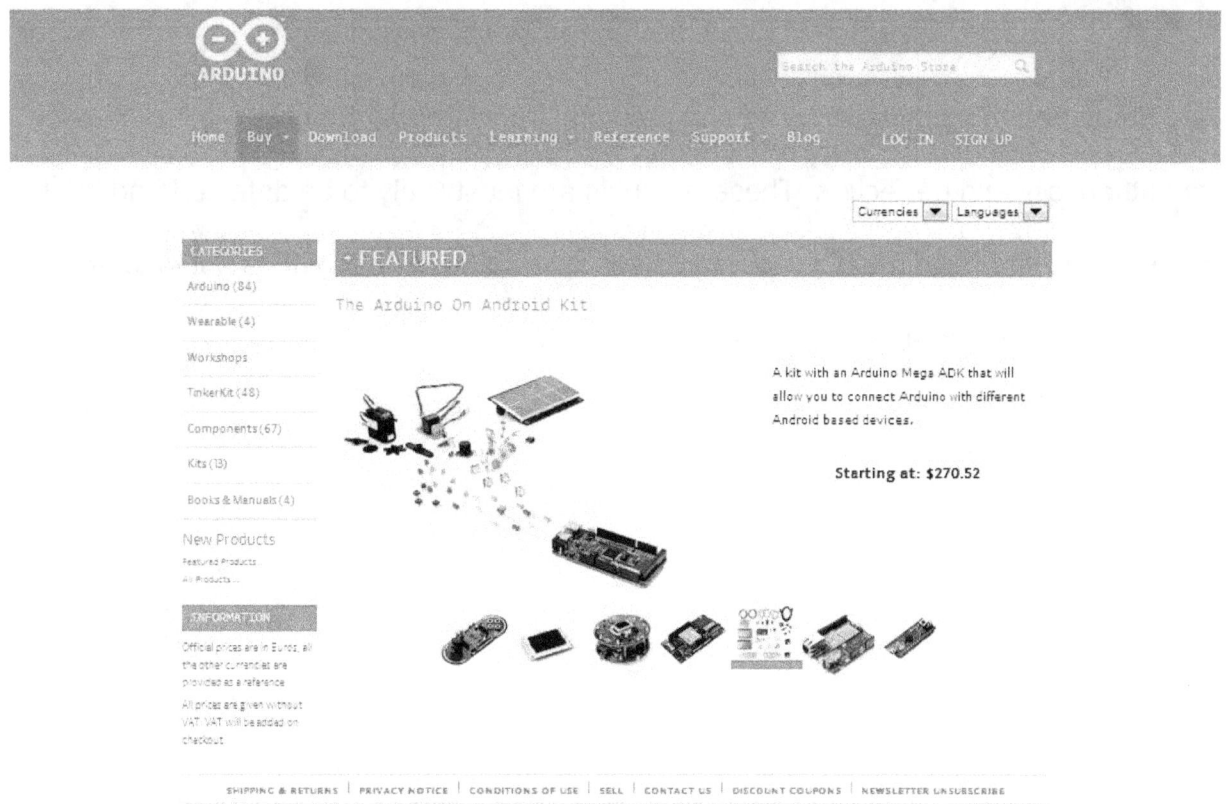

You can place an order for your Arduino board online at the Arduino Store on their official website.

The Arduino Store offers delivery options in three different geographical areas: European Union, Italy and the rest of the world.

The prices for each Arduino product are mentioned. You can change the currency for the tab placed in the top right corner. Similarly, you can change the language of the text displayed which the adjacent to the previous drop-down menu.

This mode of purchase has its own perks and disadvantages. For instance, the time elapsed between the placement of the order and the delivery of Arduino products may vary according to the location. Also, value-addition taxes may be applied on the Arduino products.

On the other hand, purchasing Arduino products from Arduino Store is a guarantee for its authenticity and perfection. These products are least likely to be defected and will be prepared according to the high standards set by Arduino. Arduino does not take any responsibility for third-party produced products.

Distributors

The market for Arduino users – prospective and old – is immense and is spread out across the world. It is not possible for Arduino to produce and deliver Arduino products everywhere in the world. As a result, Arduino products are sold via distributors' channel as well. In some areas, third-party production rights have also been provided to facilitate the trade of Arduino products.

As with the Arduino Store, this medium of trade has its own advantages and disadvantages.

The biggest advantage for this mode of sale is that it makes Arduino products accessible to you in the most affordable prices. Buying local produce is cheaper than acquiring it from another country – the shipping and inter-country trade rules govern this exchange, accentuating the price of the product.

The biggest disadvantage of purchasing Arduino products from distributors, especially the third-party produced products, is that the parent copyrighted company is not liable for any damage or low-quality products. You cannot claim any compensation for damages or substandard products.

Choose your medium of purchase by evaluating all available options. If budget constraint is an issue, it will be best if you take the risk with local products instead of buying extremely expensive foreign goods. However, if you are an Arduino enthusiast, it is highly likely the local produce will not be according to your aesthetics. Figure out your limitations before you purchase your Arduino – it can save you from unnecessary heartbreaks and difficulties. With so many alternatives, there is no reason why you cannot experiment with your Arduino projects and explore the reckless possibilities attached with it!

Conclusion

To conclude it all, it is worth repeating – the Arduino products have a large store of unexplored potential. Be the first one to discover what your Arduino can do for you.

Although it is slightly technical, mastering the Arduino is not impossible. You can start by acquiring the starter's kit and building some exemplary projects out of their manual or from this eBook. Once you have built a couple of projects, you will automatically understand the role and functions of each product and component, allowing you to build your own projects.

You can amaze your guests with wonderful creations you have made from your Arduino. You can even treat yourself with some projects to soothe your nerves. With an Arduino, anything is possible. It is the present of the futuristic robots – grab the opportunity now!